FOR PEOPLE WHO HATE PHYSICS

Michael Alan Shapiro

To Nancy, Jacki and Valerie for
their love and support

And to Aunt Linda for her
proofreading assistance

Farewell...

54

Introduction

So you hate physics? Why? Do you hate math? Do you just hate science in general? Did you have boring science and math teachers in school? That saddens me, because physics is the basis for understanding the entire physical universe and everything physical that happens within it. That's everything in the universe that is not "spiritual," if, indeed, anything "spiritual" happens in the universe!

What is the basis for the expression "What goes up must come down (which is not necessarily so)?" Why is the sky blue? Why are "thrill rides" so thrilling? What happened "in the beginning" of the universe and thereafter?

If you have any interest in these and similar questions, but don't want to learn and work with mathematical formulas to understand the answers, I promise I will teach you, *conceptually* rather than mathematically, how to understand the workings of the universe and everything in it.

As a fairly beloved high school physics teacher and author, I will make these and many more concepts easy to understand and, believe it or not, even enjoyable to learn. So, if you have the slightest interest in this stuff, this book will take you through it painlessly. Enjoy the ride!

In the Beginning

The book of Genesis states something like "In the beginning, God created Heaven and Earth." And thereafter, created man and beasts and everything else. If pinned down, I would classify myself as agnostic religion-wise. I have seen no specific evidence of God's existence, nor do I have evidence that God does not exist. Incidentally, some of the greatest scientists of all-time, including, for one, Albert Einstein, believed in God. A simplistic reconciliation of belief in God and belief in science is: God created the universe and the laws of physics. Thereafter, the universe took its course in accordance with the laws of physics.

Belief in God and belief in a literal interpretation of the Bible are not necessarily tied to each other. I'm sorry to now take issue with the Bible and perhaps lose some of my audience, but the Bible was not written by God nor dictated by God. It was written by religious scholars, beginning about the 3rd century B.C. So, in my opinion, it's not necessarily accurate in general and is surely not scientifically accurate. There, I said it. If you believe in a literal interpretation of the Bible, including "creationism", that Man was created some 5700-odd years ago, then you cannot integrate science with Bible and must reject science. If that is the case, please save yourself the fury of reading on and taking issue with everything I say. (I'm sure such people will stop reading anyway! In fact, I'm probably talking to myself here, since no creationist is picking up this book in the first place.)

In the beginning, according to the science of physics, the entire universe and everything that it's composed of was about the size of a pea. Amazing! Think about it...the matter that makes up every living thing on Earth, the Earth itself, the Sun, the planets, the billions of other stars in the Milky Way Galaxy and the billions of other galaxies that presently exist in the Universe were compacted to about the size of a pea!

If you think about it, that means that all that matter must have been more tightly packed, more dense, than is even imaginable. And it was. To boot, all that matter being so tightly packed together had a temperature of something like 10^{30} degrees! What is 10^{30} degrees, you ask? 10^{30} is a compact way of expressing the number 10 x 10 x 10 x 10 x 10 x 10 x 10 x10 x 10 x 10 x 10 x 10 x 10 x 10 x 10 x 10 x 10 x10 x 10 x 10 x 10 x 10 x 10 x 10 x 10 x 10 x 10 x10 x 10 x 10, i.e., 10 multiplied by itself 30 times, a/k/a 10 to the 30th power, or a 1 with 30 zeros after it: 1,000,000,000,000,000,000,000,000,000,000. Very large numbers and very small numbers are easier to write and easier to read if they are expressed in this way, known as "scientific notation."

As far as you may be concerned, suffice it to say that 10^{30} degrees is incredibly hot. The surface of the Sun is about 20,000 degrees (2×10^4 degrees) and the Sun's interior reaches a maximum temperature of about 10,000,000 degrees (1×10^7 degrees.) You may be asking, is that Fahrenheit or, what was that other one? -- Celsius? -- degrees or, if you ever even heard of it, degrees Kelvin? It's Fahrenheit, not that it really matters much.

Other than a 32 that must be added or subtracted from Fahrenheit to get to Celsius, or the other way around (I know which is which, but it's not important), Celsius degrees are 9/5, or 1.8, or roughly twice the size of Fahrenheit degrees. The addition or subtraction of 32 degrees becomes negligible when you are talking about degrees in the thousands, millions or more.

So, 10,000 degrees Celsius is very nearly, plus or minus 32 measly degrees, 18,000 degrees Fahrenheit. And the difference between 10,000 and 18,000 isn't even so important; suffice it to say the surface of the Sun is way too hot for comfort, to say the least; it would melt and then vaporize anything that approached it; and the interior of the Sun, at 10,000,000 degrees, is truly incredibly hot; hot beyond our imagination, since the hottest thing we've ever experienced is a pizza oven, at about 600 degrees Fahrenheit or an open flame, ranging from about 200 to 2500 degrees Fahrenheit from the red portion (outer edge) to the blue portion (center).

Hey, you might be thinking, didn't I say I would spare you of math? As much as possible, I will. And I *promise* you I will spare you of mathematical formulas, like $F = ma$, $p = mv$, $F_g = Gm_1m_2/r^2$ and the dozens of others that physics students must learn. The only formula I will address in this book is Einstein's famous formula $E = mc^2$, because it is so famous and so important and, to laypeople, the essence of what they believe physics to be.

So, in the beginning, the entire mass of the universe was compressed into a tiny portion of space — a tiny volume — and was incredibly hot. In fact, space itself was only the size of a pea. By the way, this beginning I speak of occurred approximately 13.7 billion years ago.

How, you may ask, do we know what the universe was like 13.7 billion years ago? Good question. The answer is: there are two basic ways, both of which agree with each other. One way is based upon interpreting the "Doppler red shift" of light from distant galaxies. Knowing what wavelengths of light are emitted by nearby galaxies, when we look at distant galaxies, we see a "red shift" of their light — the light from these galaxies is redder than the light from nearby galaxies— which occurs when a light source, in this case, a galaxy, is moving away from us.

It turns out that virtually all distant galaxies are moving away from us, and us from them, and the farther away they are the faster they are moving. In other words, the universe is expanding, like a balloon that is being blown up. And, the farther away the galaxy, the greater the red shift — the faster it is moving away from us. The fact that the universe is expanding in all directions leads us to the belief that, in the beginning, the matter of the universe was all in the same place and that some event occurred to cause all the matter to move away in all directions — the Big Bang.

By the way, despite its name, the Big Bang was not an explosion. The name was given to the event by a nonbeliever at the time, noted astronomer Fred Hoyle, who cast aspersions on the theory by giving it that name. As it was, the name stuck, even though we have come to understand that the universe did not explode into its current state but rather, for some reason and by some mechanism still unknown to us, suddenly underwent rapid expansion.

By measuring red shifts of distant galaxies, we can calculate how far away the most distant galaxies are from us and how fast they are moving away. With this data, we calculate that the universe has been expanding for 13.7 billion years.

So, one set of evidence of the Big Bang is the Doppler red shift of distant galaxies. The second piece of evidence of the Big Bang is the 2.7 degree cosmic microwave background radiation in the universe. The Big Bang Theory (the scientific theory, not the TV program) predicted that there would be a leftover heat signature in the universe from the original heat of the the pea-sized universe. That heat signature was found, accidentally, by two scientists at Bell Laboratories in New Jersey in 1965.

Arno Penzias and Robert Wilson consistently suffered interference when they aimed Bell Labs' new radio telescope in various directions. After taking every possible step to verify that the interference was not due to any malfunction of the radio telescope, they realized that the interference was coming from space itself; that all of space, in every direction, had a background microwave radiation that matter at a temperature of 2.7 degrees above absolute zero (2.7 degrees Kelvin) would radiate.

Temperature is actually a measure of the kinetic energy — energy of motion — of atoms or molecules.

Even in a solid, frozen state the molecules that make up matter move and vibrate, or, you might say, jiggle. Absolute zero is the temperature at which everything would be frozen so solid that not even the molecules that make it up would be able to move or vibrate.

Thus, Penzias and Wilson discovered the background temperature of the universe due to the Big Bang. This discovery was such a strong confirmation of the Big Bang Theory that even Fred Hoyle, who initially scoffed at the theory and gave it its name in jest, now accepted it. Penzias and Wilson won a Nobel Prize in Physics, not for their accidental discovery of some interference from space, but for their realization that what they had found was the missing link that would remove all doubt about the validity of the Big Bang Theory.

And Then What Happened?

At the time of the Big Bang, the universe was so hot that matter, in the forms that we know it today, could not exist. As I wrote above, heat is kinetic energy, energy of motion. At the time of the Big Bang, it was so hot that what we today call subatomic particles of matter — the protons, neutrons and electrons you may have come to hate in high school chemistry, and a host of others — were traveling too fast to be able to combine and remain stable. In other words, atoms could not exist. In fact, protons and neutrons could not even exist, as they are composed of smaller particles, quarks, which would have been moving too fast to come together and remain stable.

In short order, though, due to expansion and cooling of the universe, quarks were able to form protons, neutrons and other particles, and combine to form atoms of hydrogen and helium, elements 1 and 2 on the periodic table. Finally, beginning 150 million years after the Big Bang, large-scale structures — stars and galaxies — were formed as gravity caused matter to clump together. All other naturally occurring elements, from lithium through uranium (elements 3 through 92), were created by stars.

From 150 million years post Big Bang to date, the universe has continued expanding and cooling and operating in accordance with the laws of physics established about 1/1000 of a second after the Big Bang.

As we understand physics today, there are four fundamental forces of nature: gravity, electromagnetism, the so-called strong nuclear force and the so-called weak nuclear force. At the time of the Big Bang, only one force of nature existed, called the Super Force, and, as time proceeded and the universe expanded and cooled, the other three forces sprouted from the initial Super Force. Again, the metamorphosis of the Super Force into the four forces we know today occurred in a short time, within about 1/1000 second after the Big Bang.

Amazingly, physicists believe they understand the workings of the universe all the way back to 10^{-43} seconds after the Big Bang, when only the Super Force existed. 10^{-43} is one divided by 10^{43}. Since 10^{43} is a tremendously huge number, one divided by 10^{43} is a tremendously small number. Thus, physicists believe they understand the workings of the universe almost back to the beginning of time.

By the way, what happened before the Big Bang, if anything, is unknowable to us. Were there previous universes before this one? And are there "parallel universes" — other universes that we cannot see or detect? We don't know and don't believe we can know.

Although the universe is 13.7 billion years old, our Sun and its solar system, including the Earth, is "only" 4.6 billion years old. Stars actually go through a life cycle, whose length depends upon the size of the star. Our Sun is a very average star, whose life expectancy is about 13 billion years. Thus, we have nothing to fear; the Sun will not explode, fizzle out or even significantly vary the amount of energy it emits for about the next 8 billion years.

The larger the star, the shorter-lived it is. Very large stars explode in supernovas and cast themselves outward in all directions. Average stars like our Sun expand near the end of their lives, become red giants, and then contract and become dwarf stars. In its red giant phase, the Sun will expand from its current radius of 432,000 miles out to about Earth's orbit, 93 million miles from the Sun. In other words, it will have about 200 times its current radius and apparent size, and, therefore, about 8 million times its current volume. But again, this won't happen for billions of years, so don't worry.

The Sun, the Earth and everything in the solar system was formed from matter generated by stars or spewed out in the supernova death of one or more very massive stars. We are all star material, literally!

Forces of Nature

At or immediately after the Big Bang, there was one force of nature, which we call the Super Force. In fairly short order, the four forces of nature known today were borne out of the Super Force. Those four forces are: gravity, electromagnetism, the strong nuclear force and the weak nuclear force.

What is a force? A rather meek definition, yet the best we can muster, is that a force is a push or a pull. A force pushes or pulls on material objects. In everyday life, we encounter and produce many forces. Every time you move you exert a force. Every time you lift something or move something you exert a force. If you hit or kick a ball, you exert a force. If somebody bumps into you they exert a force on you — and, by the way, you exert an equal force on them.

All of the above examples of forces are not forces of nature; they are forces produced by man or animal or machine. And, it may seem obvious, men and animals and machines must have contact with the thing they are to move or lift to be able to apply a force to it.

Forces of nature are different in that regard. Forces of nature are "non-contact" forces — they occur whether or not there is contact with the object upon which the force is applied.

Think about it: if you jump up from the Earth, the Earth pulls you down via a force of gravity. You are not then in contact with the Earth — you are in midair

— yet, the Earth applies a downward force on you. Or, if you have a magnet at your disposal and want to lift a paper clip, you do not have to actually touch the paper clip with the magnet. When you bring the magnet close to the paper clip, it will jump up to the magnet. Try it!

So, there are four different forces of nature that act without the need for contact: gravity, electromagnetism, the strong nuclear force and the weak nuclear force.

Gravity

You are certainly somewhat familiar with the force of gravity. The force of gravity exerted by the Earth keeps us on or near the Earth. When we throw something upward, it inevitably comes down. By the way, I stated in the introduction that what goes up does not necessarily come down. If you were able to throw a ball upward with a speed of about 17,000 miles per hour, the ball would escape Earth's gravity and keep on traveling away from the Earth forever. And, equally as likely as you being able to throw a ball at this speed, pigs can fly.

The force of Earth's gravity on us is what we refer to as our weight. Would you like to lose 83% of your weight in three days? You can do it! Just travel to the moon, and bring a scale. The same scale you used on Earth to measure your weight will read 1/6 of what it reads on Earth. The moon's gravity is very nearly 1/6 as strong as Earth's gravity. And although you would "lose" 5/6 of your weight simply by being on the moon versus being on Earth, unfortunately you would not be any thinner. You would still have the same mass; you would still be composed of the same amount of matter. What you really want to lose if you feel you are "overweight" is mass; you are not "overweight," you are "over-massed."

So, gravity is an ever-present force that pulls us down to Earth if we are in the air and pretty much keeps us tied to the Earth. It took thousands of years of civilization for us to create technology powerful enough to escape Earth's gravity — space ships. Airplanes are powerful enough to allow us to very partially escape Earth's gravity and hover a relatively short distance from the surface of the Earth. The Earth's radius is about 4000 miles, and commercial jetliners reach an altitude of a mere 7 miles. Even the International Space Station, reached by space shuttles, is only 205 - 270 miles above the surface of the Earth. That's merely the distance from New York City to Washington DC.

You may believe that the Earth is the only thing that exerts a force of gravity. If so, I must say you are wrong. The moon exerts a force of gravity that is 1/6 as strong as Earth's gravity. Jupiter exerts a force of gravity about 3 times stronger than Earth's. In fact, every material object in the universe — galaxies, stars, planets, asteroids, as well as people, animals, trees, grasshoppers, protons, neutrons and electrons — exert a force of gravity on every other material object in the universe. This is the essence of Newton's Universal Law of Gravitation: everything in the universe gravitationally attracts every other thing in the universe.

So why are we generally unable to jump beyond a few feet in height from the Earth, yet it is a simple matter to walk away from another person, or a tree, or a grasshopper? The answer is: Earth's gravity is much, much stronger than the gravity exerted by a person, a tree or a grasshopper. Why would that be? If you think about it, I believe you will come up with the answer. Are you thinking? What's your answer? Is your answer that the Earth is much more massive than a person, a tree or a grasshopper, and therefore its force of gravity is much stronger? If so, you are correct. The force of gravity exerted by an object is directly proportional to its mass. The greater the mass of an object, the greater its gravity.

In fact, gravity is the weakest force of nature, by far. For its force of gravity to be significant, an object must be tremendously massive, like a star, a planet, or a large moon. While the Earth exerts a force of 100 to 300 pounds, more or less, on any given adult person, the gravitational attraction between two consenting adults is about 10^{-7} pounds, i.e., about 0.0000001 pound. In other words, the gravitational force between any two everyday objects is negligible.

If gravity is such a weak force, how can it hold our bodies together, or hold together the material that forms everyday objects, like tables, chairs, trees, and, beyond that, hold them together so tightly that it takes great force to break such objects? The answer is...it's not the force of gravity that holds us and everyday objects together; it's electromagnetism.

Electromagnetism

Electromagnetism, as you might imagine, is a combination of the forces of electricity and magnetism. What is electricity? Static electricity is a buildup of electric charges and current electricity or electric current is the flow of electric charges.

Static electricity can build up in clouds and, when it discharges and flows, lightning is produced, with thunder as a byproduct. As you have also probably witnessed, static electricity can build up through friction when you walk on a carpet or pull a wool sweater over your head. The discharge of that buildup of excess electrons actually produces the same phenomenon as lightning, on a smaller scale, and causes you to feel an electric shock when the excess charge is discharged. Try dragging your feet on a carpet and then pointing your finger towards a metal doorknob from a very close distance (about one centimeter, or half an inch, or less). You will discharge your excess static electricity to the doorknob, and, in the process, a tiny bolt of "lightning" will be produced. And you will feel a mild pain of electric shock.

While static electricity is basically an unwanted inconvenience in our lives, electric current is an extremely desirable phenomenon. Electric charges commonly flow in wires or, more commonly these days, in solid state circuitry, a/k/a "chips." Chips embedded in calculators, computers, iPhones, iPads, Gameboys, TVs, cars and just about all of our current technology control the function and operation of these technologies. Tiny chips are the equivalent of the mass of tangled wires that carried electricity in "the olden days" — up to about 1970 or so, for the general public, and about 1950 for scientific applications.

As you might imagine, electricity is a key aspect of modern life. Virtually all of our technology is electricity- or electromagnetic-based. The list is endless: computers, televisions, iPhones, iPads, automobile systems, lighting, heating and air conditioning, and just about every device you use on a daily basis; not to mention the devices we rely on outside of our own household: traffic lights, power grids, water delivery systems and the computers that control these systems.

If, hypothetically, there was some kind of huge electromagnetic burst, either from space or from a new weapon system, that could knock out the computer systems of the U.S.A., we would be paralyzed. Banking and finance records destroyed or at least unavailable, systems for delivery of goods and services halted, communications suspended. We would effectively be living in the 18th century, without the adaptive abilities of the people of that time. Farmers would have food. People with wells would have water. Everybody else would be "SOL". Mass hysteria and lawlessness would ensue as everybody who was living smoothly and comfortably suddenly had to use any available means to obtain food and water.

To me, this is the scariest scenario of modern life. However, the good news is that a huge electromagnetic burst from space is highly unlikely.

The Sun's electromagnetic outbursts are very consistent, and, indeed, essential to our existence. And there do not appear to be any very massive stars that could explode in a supernova close enough to us to do significant damage.

The bad news is that our enemies or potential enemies are working to develop weaponry or other means to cripple our computer systems and networks. We don't hear about it much, but the good ol' U.S.A. is also working to develop such weaponry and means (I imagine we already have developed such things), as well as to develop defenses to such things.

We have recently accused China and Russia of hacking US government and corporate computer systems, and Germany has accused us of hacking Angela Merkel's emails. So far, such hacking has not risen to weaponry, but the threat is palpable.

Aside from providing us the tools for easy living in the 21st century, you might say that electricity is the most important force in nature for another, more crucial reason: it is the force that holds our bodies together and upright and facilitates all of our life-sustaining biological processes.

Remember high school chemistry? Remember the key ideas of chemistry? The concept of valence *electrons* being exchanged or shared in chemical reactions, and the concept of *electro*negativity? Simply stated, all chemical reactions, including those that sustain us, are electrical. Covalent bonds, ionic bonds and intermolecular forces are all pseudonyms used by chemists for electrical forces.

If electrical forces ceased working in our bodies, we would turn into a mass of goo that would fall to the ground, and, incidentally, we would die. Gravity is way too weak to hold our bodies together. The electrical force between a proton and an electron is an astounding 10^{39} times greater than the gravitational force between them. Or, in other words, the force of gravity is negligible in and between everyday objects, plants, animals and people; it is electrical forces that keep us alive and functioning.

So, that's a thing or two about electricity. What about magnetism? You know that a magnet "sticks" to and attracts metal things. Actually, it only sticks to and attracts certain metals: primarily iron, nickel and cobalt, and alloys of these metals, including steel. Aluminum, a very common metal, is not magnetic. The magnetic properties of iron, nickel and cobalt are due to the fact that the electrons of these metals all spin with a common axis. Other metals and other elements' electrons' spins are not aligned; they are random, so the magnetism produced by all their individual spins cancel and produce no net magnetism.

Thus, magnetism, which is a product of electron spin, is intimately connected to electricity. Although a magnet sticking to a refrigerator and electricity flowing in your iPod to produce music seem like two completely different phenomena, they are in fact two sides of the same coin; two facets of the same exact force.

Prior to the early 1800s, electricity and magnetism were considered to be two separate and unrelated forces of nature. Then, Andre-Marie Ampere demonstrated that moving electric charges produce a magnetic field and Michael Faraday demonstrated the flip side — that a changing magnetic field produces electric current. The merger of electricity and magnetism was ultimately stated in four elegant equations by James Clerk Maxwell in the early 1860s. Additionally, Maxwell's equations infused the concept that light is an electromagnetic wave.

You may have heard of the electromagnetic spectrum. The electromagnetic spectrum is the entire collection of different electromagnetic waves. From highest to lowest energy, they are: gamma rays, X-rays, ultraviolet light, visible light, infrared radiation, microwaves and radio waves. Yes, all of these phenomena are, in essence, the same. In fact, physicists refer to all of these phenomena collectively as "light." In common parlance, we use the term "light" to mean "visible light", but, physically, all these phenomena are essentially the same. The differences among them are their frequencies, wavelengths and energies. All electromagnetic waves are produced by accelerating charged particles.

The most energetic electromagnetic waves, gamma rays, are produced in nuclear reactions. The breakup of an atomic nucleus, as, for example, in nuclear weapons, releases enormous amounts of energy. In nuclear reactions, some of the mass involved in the reaction is annihilated and turned into energy. The amount of energy produced by the annihilation of mass is given by the most famous of physics equations, $E = mc^2$.

So what does this equation mean? Symbolically, the "E" stands for energy, "m" is for mass and "c" is the speed of light. In words, the equation is: energy equals mass times the speed of light squared. As you may know, light travels extremely fast. In fact, its speed is downright astounding. The speed of light in a vacuum (empty space) is 186,282 miles per second or, for people elsewhere than in the United States, 300 million meters per second or 300,000 kilometers per second. Ironically, the United States, which revolted against England in 1776, is the only country in the world that still uses the "English system" of measurement. Even England uses the "metric system", also known as the Systeme Internationale, in the original French, or International System.

You read that correctly: light moves at a rate of 186,282 miles per *second*. In one second, light can travel 3/4 of the way to the moon or circle the Earth seven times. On the other hand, the Earth is far enough away from the Sun that sunlight takes 500 seconds, or 8-1/3 minutes, to travel to Earth. When we look at the Sun, we are seeing it as it looked and where it was located 8-1/3 minutes ago. And when we look at stars in the sky, we are seeing the light they emitted years or even centuries ago. The next nearest star to us aside from the Sun is Proxima Centauri, which is 4.3 light years from us. Its light takes 4.3 years to reach us. Many of the other stars we see are hundreds or thousands of light years from us. Our galaxy, the Milky Way, is 100,000 light years across. A light year, by the way, is not a period of time, but, rather, the distance light travels in a year: 5.88 trillion miles or 9.42 trillion kilometers.

Since the speed of light is a very large quantity, the speed of light squared is much larger still. So, the essence of the equation $E = mc^2$ is that the annihilation of a small amount of matter produces a great amount of energy. This was the basis for the atomic (really, nuclear) bombs dropped on Hiroshima and Nagasaki, and the reason the world is so concerned about mad dictators gaining access to nuclear weapons.

Annihilating matter and producing energy is also the *modus operandus* of the Sun and every other star. Every second, the Sun annihilates tons of its matter and converts it into a tremendous amount of energy.

Let's think about the energy produced by the Sun. Realize, first, that the Earth receives a very small amount of the total energy emitted from the Sun. The Sun spews out energy in all directions. The tiny Earth, 93 million miles away from the Sun, only receives a tiny slice of that energy.

Imagine the Sun radiating energy in all directions. The Earth has a surface 8000 miles wide, 93 million miles from the Sun. The area of the Earth compared to the area of an imaginary sphere 93 million miles in radius is 1/540,562,500. Thus, the Earth receives about one 540 millionth of the Sun's total radiation. Nevertheless, the energy the Earth receives from the Sun ultimately accounts for most of the energy available on Earth.

The Sun drives the Earth's oceans, produces the Earth's winds and directly provides energy to plants, which are eaten by animals to give them their energy, which are eaten by other animals to give them their energy.

But hey, you might think, much of our energy, other than energy from what we eat, comes from coal and petroleum. That's true. What is coal and petroleum? They are substances formed from decayed, compressed plants and animals, whose energy was delivered by, you guessed it, the Sun. As students learn in middle school science, if not earlier, the Sun is the ultimate source of all energy on Earth.

The Sun radiates a variety of electromagnetic waves. In fact, it radiates every type of electromagnetic wave listed above: gamma rays, X-rays, ultraviolet, visible, infrared, microwave and radio. But, as you might imagine, the greatest percentage of the Sun's electromagnetic radiation is visible light. That is why our eyes have evolved to be able to detect that form of light, which we call "visible" because we can see it. The Sun radiates a smaller amount of ultraviolet light and infrared light, and even less x-ray and radio waves.

We have spoken a bit about gamma rays, which have so much energy that they would pass right through our bodies if we were exposed to them. The next most energetic electromagnetic waves are X-rays, which, as you know, are energetic enough to penetrate our soft tissue (skin, fat, muscle and the like) but are stopped by the densest portion of our bodies, our bones. Thus, we use x-rays, in very small bursts, to look at our bones and detect fractures.

Less energetic than x-rays are ultraviolet rays, which can penetrate several layers of skin and thereby cause "sunburn." Visible light, infrared light, microwaves and radio waves lack the energy to penetrate our bodies.

But wait, you might be thinking, aren't microwaves powerful enough to cook food and, if we were exposed to them, cook us? Yes, but...the specific frequencies and wavelengths of microwaves used in microwave ovens are special because they match the "resonant frequency" of water molecules.

Molecules of every substance, including water, vibrate back and forth as if they were interconnected to each other with springs. The specific frequencies of microwaves used in microwave ovens match the frequency of water molecules' vibrations. As a result, they cause the water molecules to vibrate over greater and greater distances, thus causing the molecules to move faster and faster. And remember, heat is the kinetic energy (energy of motion) of molecules.

Long story short, microwave ovens cause water to heat up, even boil, *and do nothing else*. You cannot heat a rock in a microwave oven, because a rock has virtually no water content. You cannot heat a dinner plate in a microwave oven, because a dinner plate has virtually no water content. You can cook meat, fruit and vegetables in a microwave oven because they have a high water content.

But if you cook some food on a dinner plate in the microwave oven, doesn't the plate get hot? Yes it does, but not from the microwaves. The plate heats up by heat transferring from the hot food to the plate;

the plate heats by conduction (contact) with the hot food, not by radiation of microwaves.

Try this experiment at home: put a dinner plate without food in a microwave oven, and microwave it on high for one or two minutes. See if the plate heats up. It won't!

Visible light is very important to us, since we have developed eyes to see visible light. Technologically, however, other than for us to see objects, visible light is not very important to us. In the year 2016, microwaves and radio waves are most important to us technologically. All of our forms of communication, other than those dinosaur "land lines," rely on microwaves and radio waves. Wireless phones, the Internet, iPads, TV remote controls, garage door openers, radios, of course, and myriads of other daily conveniences use microwaves or radio waves.

The Strong and Weak Nuclear Forces

Rounding out the four forces of nature are the strong and weak nuclear forces.

An atom consists of a nucleus composed of protons and neutrons, with electrons moving outside of the nucleus. An atom is about 10^{-10} meters in radius. As small as the atom is, the nucleus is 10,000 times smaller, about 10^{-15} meters in radius.

The protons and neutrons are tightly packed in the nucleus. If you think about it, though, since protons have positive charge and neutrons have no charge — they are neutral, hence, their name — the protons all repel each other. As you may have learned in middle school science, like charges repel and opposite charges attract. So what holds nuclei together?

We have already established that gravity is far too weak to hold atomic nuclei together; the gravitational force of attraction between two protons is 10^{36} times weaker than the electrical force of repulsion between them. Thus, there must be another force holding the nucleus together. That force is the strong nuclear force.

The strong nuclear force exists only between nuclear particles — protons and neutrons and certain other forms of particles called hadrons — and, in atoms, only acts within the atomic nucleus. Being about 100 times stronger than the electrical force of repulsion between protons, it keeps protons and neutrons tightly bound together in the nucleus.

Another nuclear force, the weak nuclear force, is responsible for nuclear decay — larger, unstable nuclei morphing into smaller, more stable nuclei and producing radioactivity (gamma radiation) in the process. For example, uranium naturally, of its own accord, goes through a series of radioactive decays and ultimately, over billions of years, turns into lead. In these processes, high energy gamma rays are emitted. When exposed to significant doses of gamma rays, or their lower energy brethren, x-rays, humans and other animals are subject to a variety of diseases, including cancer, with the possibility of death. In fact, Marie and Pierre Curie, Nobel Prize winners and pioneers of radioactivity, both died from overexposure to radiation, being unaware of its danger.

Now that we've become familiar with the four forces of nature, let's understand what forces do. The man responsible for our basic understanding of what forces do is Sir Isaac Newton.

Newton's Laws of Motion

Sir Isaac Newton was perhaps the greatest physicist of all time. He, himself, would have no problem stating categorically that he was the greatest physicist of all time, since, besides being so, he was hugely self-aggrandizing and a social recluse. He pretty much spent all his time by himself, thinking about motion, gravity, light, optics and the mathematical relationships governing them.

In addition to contriving the laws of motion that explain almost every mechanical interaction, the Universal Law of Gravity and a variety of other theories in physics, he invented calculus approximately contemporaneously with Gottfried Wilhelm Leibnitz and spent a good deal of his time and energy thereafter attempting to convince the academic world that he invented calculus prior to Leibnitz and thus deserved sole credit for it.

Newton's three laws of motion can be stated as follows:

First Law (Law of Inertia): An object at rest stays at rest and an object in motion continues in motion in a straight line at constant speed unless acted upon by an outside unbalanced (net) force.

Second Law: A net force acting upon an object causes the object to accelerate in the direction of the net force.

Third Law: For every action force there is an equal and opposite reaction force.

Let's explore these laws. What are you doing right now? Sitting in a chair? Let's say you are. If so, there are basically two forces acting on you — gravity and a "normal force." Not only when you sit in a chair, but at all times, Earth's force of gravity pulls you down towards the center of the Earth. Yet, as you are sitting in the chair, you do not fall to the Earth. Why? Because the chair and the floor together exert an upward force exactly equal in magnitude to the force of gravity exerted downward on you by the Earth. The force exerted by a surface to support an object on the surface is called a "normal force." By the way, it's not called a "normal" force because the force is "usual;" it's called a "normal" force because the force acts perpendicularly to the surface. In mathematical vernacular, "normal" means "perpendicular."

So, while you're sitting in your chair, a downward force of gravity is exerted on you by the Earth and an equal upward normal force is exerted on you by the chair and/or the floor. The result is that these forces cancel each other. Thus, there is no unbalanced force acting on you; the net force on you is zero. Thus, according to Newton's First Law, you stay at rest (you do not move).

How about a car that's moving down the highway in a straight line at constant speed? According to Newton's Second Law, there must be no unbalanced forces — no net force — acting on the car. What forces are acting on the car? A downward force of gravity and an upward normal force, equal in magnitude and opposite in direction to gravity, just like you sitting in a chair. These two forces cancel each other. How about horizontal forces? There's a force of thrust exerted on the car by the engine, propelling it forward, and there are frictional forces exerted on the car in the opposite direction, impeding its motion. If the car is traveling at constant velocity, the forward thrust is exactly balanced by the backward frictional forces (a combination of air resistance, friction between the tires and the road and friction of internal parts of the car). If, however, the car is accelerating forward, the forward force of thrust is greater than the retarding frictional forces. If the car is slowing down, the frictional forces, including forces due to applying the brakes, are greater than the force of thrust, if any. Newton's Second Law; it works!

Newton's Third Law is usually the most confusing to beginning physics students and the hardest to accept. For every action force, there is an equal and opposite reaction force. What does that mean? For example, if you lightly tap on a table, it taps you back with a force of the same magnitude (size) in the opposite direction. Try it. Now hit the table a little harder. It hits you back a little harder. Why should you not hit the table *really* hard? Because it will hit you back really hard, and perhaps break your bones.

Some other examples of Newton's Third Law that students find difficult to accept are:

(1) The force of gravity the Moon exerts on the Earth is equal in magnitude to the force of gravity Earth exerts on the Moon, even though the Earth has 81 times the mass of the Moon.

(2) The force that a mosquito exerts on the windshield of a car speeding down the highway is equal to the force that the car exerts on the mosquito. Really? Yes! Try to imagine the collision of the mosquito and the car in slow motion. At first, the car and mosquito are barely touching, and the force is minimal. In the next 1/100 of a second or so, the actual contact between the mosquito and the car is greater, and the force exerted is greater, but still not sufficient to squash the mosquito. The car is exerting a small force on the mosquito and the mosquito exerts the same force back on the car, just like when you merely tapped the table. Within a few hundredths of a second after initial contact, there is full contact between the mosquito and the windshield. However, once the car exerts a force on the mosquito equal to the maximum force that the mosquito is capable of exerting on the car, the collision is over and the mosquito is squashed onto the windshield. That force, as you might imagine, is quite small and, accordingly, the motion of the car is not significantly affected by the collision with the mosquito. The motion of the mosquito, however, was greatly affected. Indeed, the mosquito's motion ceased as a result of the collision. Applying Newton's Second Law to the situation, the small force involved in the collision between the mosquito and the car resulted in a complete deceleration of the mosquito, since the mosquito has minimal mass, but resulted in negligible deceleration of the car because the car has a much greater mass. Likewise, when a speeding railroad train hits a car on the tracks, the car is rapidly accelerated to nearly the speed of the train and is destroyed, while the train is slowed a bit and not damaged as much. And while the occupants of the car will likely be killed, the occupants of the train will likely suffer non-life threatening injuries.

Modern Physics

Atomic Structure

The discipline of physics is commonly broken up into two basic areas: classical physics and modern physics. Classical physics was first hinted at by the ancient Greeks, but really developed by Galileo, Newton and other great scientists from about 1600 to 1900 A.D. Classical physics perfectly explains all macroscopic phenomena: motion of objects we can see, static electricity, electric current, magnetism and wave phenomena.

The concept that all matter is made up of individual small units called atoms originated in ancient Greece, hundreds of years B.C. What, exactly, an atom is and how it is structured remained a mystery until the year 1897 and, indeed, has not been completely clarified to date.

The ancient Greek Democritus is credited with the idea that matter can be chopped up into smaller and smaller pieces but not *ad infinitum;* that, eventually, you would arrive at atoms, the building blocks of matter. Here's an analogy: imagine a house made of bricks that are infinitely hard. The house can be decomposed into smaller and smaller pieces, until you have just a collection of bricks, the building blocks of the house. Likewise, matter is composed of atoms, which cannot be further decomposed.

Democritus' idea is partially correct: individual elements of matter can be broken into smaller and smaller pieces, until the smallest piece of an element is achieved, the atom. However, it turns out that atoms have an internal structure: they can be broken down into smaller pieces (particles): protons, neutrons and electrons.

J.J. Thompson is credited with discovering that the electron is a particle existing within that atom and having negative electric charge, in 1897.

In 1910, Ernest Rutherford discovered that the atom has a small, dense nucleus with positive charge and that the atom is mostly empty space. In 1912, Niels Bohr set forth a model of the atom wherein the negatively charged electrons orbit the positively charged atomic nucleus.

Atoms composed of only protons and electrons, especially atoms with many protons, would be unstable, causing physicists to theorize the existence of neutral particles, neutrons, existing in atomic nuclei. In 1932, James Chadwick demonstrated the existence of neutrons.

Thus, it seemed, the mystery of the atom was completely and finally solved. However, firstly, classical physics could not explain the stability of atoms beyond the simplest atom, hydrogen, and, secondly, experiments with higher and higher energy particle accelerators produced hundreds of additional elementary particles to add to our nice neat bundle of protons, neutrons and electrons. From the first conundrum quantum physics was born and from the second conundrum particle physics was born. Quantum physics explains electrons as wave phenomena, rather than as particles, and sets forth complex equations that, to date, seem to adequately explain the workings of atoms. Thus, the particles that make up matter — atoms— operate by a separate set of physical laws, quantum physics, than the material objects they make up, which operate pursuant to the laws of classical physics!

Particle Physics

As stated above, high energy particle accelerators have produced hundreds of particles in addition to the well-known proton, neutron and electron we all learned of in chemistry. After decades of experimental physicists finding new particles by the dozens in the 1930s, 40s, 50s and 60s, in 1968 Murray Gell-Mann devised a scheme to classify and understand what was by then referred to as the "particle zoo". His scheme came to be known as The Standard Model of Particle Physics. According to the Standard Model, protons and neutrons (and others in a class of particles called hadrons) are composed of smaller particles called quarks. There are six different quarks, fancifully named up, down, top, bottom, charm and strange. Thus, protons and neutrons are no longer elementary; they are made up of smaller particles. Protons and neutrons are baryons, made up of three quarks. As of now, we believe the quark is an elementary particle, i.e., it cannot be broken down into smaller particles. The electron is still believed to be elementary; it is not made up of smaller particles. The electron is one of six particles called leptons.

As you might imagine, since there are six different quarks and any three can build a baryon, there are 6 x 5 x 4 different combinations of quarks and thus 120 different possible baryons. Also, there is a class of particles called mesons that are composed of a quark and an antiquark. Naturally, there are dozens of combinations of quarks and antiquarks that can compose a meson. Ergo, we have a particle "zoo", the hundreds of different particles being observed in high energy particle accelerators.

For every particle there is an "antiparticle", having all the same properties as its associated particle except one — usually, opposite electric charge. Thus, there is such a thing as antimatter, made up of antiprotons, antineutrons and antielectrons (known as positrons). When matter and antimatter meet, they annihilate each other and produce energy according to

Einstein's equation $E = mc^2$. So where is all this antimatter? It is believed that shortly after the Big Bang, both matter and antimatter existed, but there was more matter than antimatter. Thus, all the antimatter was annihilated and our universe consists almost exclusively of matter. Antimatter can be created in small quantities in high energy particle accelerators, but only exists momentarily, until it comes in contact with matter.

Another tenet of the Standard Model is that forces are carried by particles. Gravity is carried by gravitons, electromagnetism is carried by photons (light particles), the strong nuclear force is carried by gluons and the weak nuclear force is carried by bosons. Furthermore, these force carrier particles move at the speed of light. So, for example, if the Sun were to suddenly disappear, the Earth would continue to revolve around the point where the Sun was for 8-1/3 minutes, as it takes that long for light (and gravity) from the Sun to reach the Earth.

Relativity

Although Albert Einstein won the Nobel Prize in Physics in 1921 for his explanation of a phenomenon known as the "photoelectric effect", he is best known for his theories of relativity: Special Relativity and General Relativity.

The cornerstone of Einstein's Special Theory of Relativity is that light travels at one certain speed, c, which is 300 million meters per second or approximately 186,000 miles per second, relative to any observer.

Let's compare this with the classical (Newtonian) theory of relativity. Imagine you are standing on a train station platform as a train passes by at a speed of 30 mph. Imagine further that a passenger on the train is walking forward at a speed of 5 mph. You would observe the train moving at 30 mph and, if you focused only on the passenger, he/she would be moving at a speed of 35 mph relative to you on the platform.

Einstein realized that this notion of classical relativity does not apply to light. If, hypothetically, the train in the above example was moving at the speed of light (which, we will see, is impossible) and a passenger shined a flashlight in the direction of motion of the train, you, an observer on the train platform, would not perceive the light from the flashlight traveling at two times the speed of light, but still simply at the speed of light.

This sounds innocuous enough, but, when Einstein continues his analysis, the following seemingly incredible phenomena occur: mass changes as a function of speed, size changes as a function of speed and time changes as a function of speed. The faster you travel, the greater your mass. Thus, no material object — no object with mass — can travel at the speed of light, as its mass would become infinite and thereby need an infinite amount of energy to accelerate it to the speed of light. The faster you travel, the shorter you become in the direction of travel. And the faster you travel, the slower time beats!

The standard example of special relativity is the so-called “twin paradox”: one twin travels in space at high speed for years, while the other twin remains on Earth. When the twin from space returns, he will have aged less than the twin that remained on Earth.

The differences between Newtonian and Einsteinian relativity are minuscule at common speeds — the speeds of people, cars, trains and even airplanes. But, at the speeds of spacecraft and satellites, special relativity becomes significant; indeed, GPS systems, which are satellite-based, take special relativity into account in their calculations.

Another key concept that comes out of special relativity is the famous equation $E = mc^2$, which we have already discussed: a small amount of matter can be annihilated and converted into a tremendous amount of energy. As we mentioned earlier, this is the basis of all nuclear reactions, including creation of energy by the Sun and all other stars, the basis for atomic and nuclear bombs and the basis for creation of energy by nuclear power plants.

Einstein’s General Theory of Relativity, simply put, states that space and time together comprise a four dimensional universe and that gravity causes curvatures of space.

We are well aware of the three dimensions of space: length, width and height. Every physical object is three dimensional. And we commonly think of time as something completely separate and different from space. But Einstein realized that space and time are inseparable and that, together, space and time make up a four dimensional universe.

Furthermore, Einstein was able to explain gravity as a curvature of space-time rather than as a force, as Newton explained it.

Einstein's theories have so far stood the test of time and experimentation. Stars have been shown to curve spacetime and thus act as "lenses" to focus light coming from behind them during eclipses. And time has been shown to beat slower on spaceships and satellites than on Earth.

Two other phenomena you may have heard of stem from Einstein's General Relativity: black holes and worm holes. A black hole, created by one or more collapsed stars, is a region of space where gravity is so strong that not even light can escape from it. In terms of General Relativity language, a black hole is literally a hole in spacetime.

Very much related to this, it is theorized that, if you traveled through a black hole and managed to come out the other side unscathed, a most unlikely event, you would find yourself in a very different part of the universe or possibly in another "parallel universe". Since black holes curve spacetime to an extreme, you can theoretically travel from one distant point to another in space by taking a much shorter path through a black hole. In this context, black holes are known as "worm holes".

Conclusion

In the 17th, 18th and 19th centuries, it must have appeared to physicists that they had fully and completely come to understand motion, forces, matter, energy, waves and light. But discoveries in the 20th century and continuing to date have produced a schism between the neatly packaged explanations of classical physics, which still do well explaining everyday phenomena, and the concepts of quantum mechanics, particle physics and relativity that fly in the face of classical physics while better explaining atomic and subatomic physics and gravity.

The "holy grail" of physics at the moment is the reconciliation of classical and modern physics, which are completely at odds with each other. In a sense, "the more we know, the more we don't know." Will there ever be a reconciliation of classical physics and modern physics; i.e., a completely new theory of matter and energy that explains all physical phenomena? Can we and will we ever fully understand "the mind of God" in his formulation of the laws of physics? I guess we'll see…

Miscellany

In the introduction to this book, I promised that I would answer several questions. Some of them did not fit easily into the flow of the book. Since I promised I would, I will answer them now.

Why is the sky blue?

The Sun produces light of every color in the visible range of the electromagnetic spectrum, as well as invisible light: x-rays, ultraviolet, infrared, microwave and radio waves. When all colors of visible light are taken together, they appear white. Thus, we receive white light from the Sun.

You may have had some experience with a prism, usually a triangular shaped piece of glass. When white light travels through a prism, it separates into all of its component colors; all the colors of a rainbow. This is ultimately because different colors of light, comprised of different wavelengths and frequencies, travel at different speeds through a medium (e.g., glass, water, oil, diamond…) and are therefore deflected at different angles by the medium.

Well, the Earth's atmosphere itself is a medium and, accordingly, acts like a prism, breaking up white light into all its component colors. OK, then why doesn't the sky look like a rainbow all the time? Because the atmosphere (i.e., the various gas molecules in the atmosphere) absorb and re-emit light. It is essentially an accident of both what the atmosphere is composed of and how thick it is that, midday, most colors of light are absorbed by the atmosphere and a plurality of blue light is emitted. Thus, if the sky is clear, midday the sky appears blue.

And why is the sky pink, red and purple at sunrise and sunset? At midday the Sun is more directly overhead and light travels more directly through the atmosphere, thus taking the shortest path to Earth. At sunrise and sunset, the Sun is at a glancing angle to the Earth's atmosphere and, accordingly, sunlight has to travel a longer path through the atmosphere to reach Earth's surface. Again, by accident, due to this different thickness of Earth's atmosphere and its composition, a plurality of red and violet light break through the atmosphere.

And, while we're on the subject of colors in the sky, when and why do we see rainbows? Rainbows are produced by water droplets acting as prisms. There is always some amount of water in the atmosphere, as long as the relative humidity is greater than 0%. But, in general, the water in the atmosphere is water vapor, a gas, consisting of extremely small particles that do not produce the prismatic effect of a rainbow. By the same token, large drops of liquid water, which pour down on us during a rainstorm, do not produce a pronounced prismatic effect. Sometimes, only sometimes, at the tail end of a storm, liquid water droplets still in the sky are just the right (small) size to produce a rainbow. Usually within minutes, these small droplets evaporate, becoming water vapor, and the rainbow is gone.

Why are “thrill rides” so thrilling?

The answer is simple: when you feel a force, or combination of forces, unequal to your weight, it is very unusual and you feel different and possibly fearful or, if you are adventurous, trilled.

What are you doing right now? Probably sitting. What else do you do, minute after minute, day after day? You stand, you walk, you lay down and sleep. If you lead an active lifestyle, perhaps, in addition, you run, you lift weights, etc.

Let’s examine the “activities” of sitting, standing and sleeping. What forces are acting upon you when you are sedentary? The ever-present downward force of gravity (i.e., your weight) and an equal upward force supplied by whatever surface(s) are supporting your weight — a chair, the floor, a sofa, a bed.

Importantly, you only feel one of the two forces acting on you as you sit. You feel the force exerted by the chair and floor, but you do not feel the force of gravity. Gravity is a force of nature and is a non-contact force; you do not have to be in contact with the Earth to be influenced by its gravity. If you jump up in the air, the Earth pulls you down without being in contact with you. Furthermore, you are probably seldom in contact with the Earth; right now, for example, I am sitting on a chair in my study. I am a few feet above the Earth, as I walked up several steps to enter my house. You are truly only in contact with the Earth when you are on the ground (or in a body of water).

And even then, if you are wearing shoes, you are not truly in contact with the Earth.

We humans, and animals for that matter, only feel "contact forces" — forces exerted directly upon us by things, whether animate or inanimate. We feel forces exerted on us by chairs, floors and beds; by other people, by animals, and by other things we may come in contact with. When we are sedentary, the only forces we feel are the forces exerted by surfaces supporting our weight, which together are equal to our weight. (As I am sitting writing this now, a chair is supporting most of my weight and the floor is supporting a small portion of my weight.)

We are accustomed to feeling contact forces totaling our weight, as that is what we feel almost all the time. Indeed, when we feel that much contact force, we may be unaware of it, since it is so customary. When we feel one or more forces that, combined, are unequal to our weight, our reaction to this unusual situation is either fear or thrills, depending upon our disposition towards new and unusual experiences, to action and adventure.

That's the basis of "thrill rides" — subjecting us to varying contact forces unequal to our weight.

Let's examine the forces we feel on a roller coaster. You get into a seat on a stationary roller coaster, and you are subjected to the usual contact forces of sitting in a seat. Other than the anticipation of what is going to happen (emotional feelings), physically you feel perfectly normal. Now the roller coaster starts to move; it accelerates. You feel an initial jolt; you feel the back of the seat pushing harder on you than when you were at rest. OK, no big deal, you feel that frequently enough when you accelerate in a car.

Now the roller coaster starts climbing a hill. You may experience the emotional feelings of fear of heights and, again, fear of the unknown (what's going to happen on this ride), and, physically, you continue to feel excess contact force from the back of your seat. Still, no big deal physically, but, again, emotionally, fear and/or excitement.

The roller coaster reaches the top of the hill and starts down a steep incline. Now we're talking! The greater the slope, the more it "feels" like free-fall. If you were to actually experience free fall, — for example, skydiving — you would physically feel *nothing!* Remember, you only physically feel contact forces. When you're free falling, the only force acting on you is gravity, a non-contact force, so you physically feel nothing. That's scary, since you're used to feeling that your weight is being supported by one or more surfaces. (By the way, even when you're skydiving you're not truly in free fall; there is some amount of air resistance, an upward force of friction as you are literally colliding with billions upon billions (really, billions times billions times millions) of air particles. Remember Avogadro's number from chemistry? One mole of air, which is roughly one cubic foot of air, contains 6.02×10^{23} air molecules.)

OK, back to the roller coaster: as you are moving down the incline, the contact force you experience is equal to your weight times the cosine of the angle of the incline. For a 60 degree incline, that's equal to half your weight. Accordingly, you physically feel halfway between normal and free-fall; you are in semi-free-fall.

Now, you reach the bottom of the incline and head upward again. The seat has to exert a force greater than your weight to change your acceleration from downward to upward, and you feel an upward push from the seat greater than your weight.

Next, let's explore traveling around a curve. You've undoubtedly experienced this driving in a car. When you round a curve, there's a force provided to the car by friction towards the center of the circle that you're traveling in. That's called a centripetal force. What you as a driver or passenger feel is an imaginary "centrifugal" force, as if you're being pushed outward from the center of the circle. What's actually happening is that the force of friction acts on the car, not on you. You have inertia, and keep traveling in a straight line as the car turns; thus, you feel like you are being pushed in the opposite direction of the turn. On a roller coaster, the turns are much sharper than you normally experience driving, so the force you feel is much greater.

Finally, let's explore a loop-the-loop. At the bottom of the loop, you feel an upward force greater than your weight, as we discussed previously when you're at the bottom of an incline and heading up. At the top of the loop, you feel a force from the seat that's less than your weight; you feel partially or nearly "weightless", as you do in free-fall. (By the way, you are never weightless when you're on or near the Earth. At all times on a roller coaster or during skydiving, the Earth is exerting a downward force of gravity on you — that force is the definition of weight. The term “weightlessness” is a misnomer used to mean the absence of feeling any contact forces. You would truly be nearly weightless in outer space, where you are far away from any massive body and the force of gravity upon you is minimal.) When you’re halfway between the top and bottom of the loop, you actually feel a force that’s more or less equal to your weight, depending on the speed of the coaster and the radius of the loop.

So, the “thrill” of thrill rides is somewhat provided by constantly being subjected to (physical) contact forces unequal to your weight and, emotionally, both by your reaction to the different physical forces you’re subjected to and the anticipation of what might happen on the ride.

I don’t know about other people, but in my experience I’ve actually been less frightened by thrill rides that occur in the dark (like Space Mountain) or if I happened to close my eyes. Personally, I think most of my fear of thrill rides is the fear of falling out of the seat and fear of heights. In the dark or with my eyes closed, the fear of what I am seeing (a big drop) is eradicated, and I just react specifically to the purely physical forces involved. For me, it’s less scary (and therefore less “thrilling”) that way.

Farewell…

Well, that's it! I hope you enjoyed the ride, and I hope you experience a greater richness in your life with your greater understanding of the phenomena that occur to you and around you. Farewell!

Printed in Great Britain
by Amazon